The Answer

Is You

The Formula to Master Your Mind and Your Emotions

Mimi Bland

WOW Book Publishing™

'There are only two ways to live your life. One is as though nothing is a miracle. The other is as though everything is a miracle'.

— Albert Einstein

Dedication

I wrote this book to teach you the power of your thoughts. What is the power of your thoughts? The power of your thoughts is . . . what you envision in your mind and create in your thoughts, you can manifest in your life. By accepting that you can redesign any aspect of your life through your thoughts, you will come to realise that you alone are the sole creator of your destiny. May this book offer you an abundance of peace, love and happiness. May it guide you through the journey of life as you create a life that you love to live, each and every day.

I wish to dedicate this book to my mother, in memory of her constant love and support, and in gratitude to her for her influence on my life journey.

I love you, Mum x

Contents

Foreword

Dear Reader,

The Answer is You is the book you need to read when you are lost and searching for your purpose in life. Once you start reading, you will need to apply the knowledge and information provided to you in the book so that you can begin the process of transforming your life. Mimi has acquired some masterful skills and knowledge, and she imparts them to you in a way that will allow you to understand and apply to make positive changes in your life immediately.

The knowledge in this book has the power to create the life you truly deserve and desire.

I can tell you from my own personal experience that Mimi has the expertise, skills, spirit, and heart necessary for helping you make the positive changes in your life so you can live a happy and fulfilled life.

I invite you to make this book a part of your life. You are most definitely worth it.

— Vishal Morjaria
Award-Winning Author and International Speaker

'Learning to love yourself Is
the greatest love of all'.

Dear Reader

Mimi wrote this book to inspire as many people as possible to take complete control of their life experiences.

During her life, Mimi faced several challenges, both mentally and emotionally, which she overcame.

Mimi has a gift for identifying and seeing the true, inner qualities of people. This allows her to connect both spiritually and vibrationally with them to complement their personal mindset in order for them to achieve life-changing transformations.

She has the gift of seeing the best in people, but more importantly, she is able to show them how to feel, see and realise their true potential.

Mimi is a visionary who has a very strong belief system that aids her in offering a service that leads, inspires, motivates, and empowers people to build self-esteem and confidence.

She has a great gift in human understanding complemented by over 20 years' experience in training and gaining knowledge in the fields of mindset, positivity, clarity and spirituality.

She is here for one reason, and that is to coach and teach the power of the mind by promoting the ability to understand 'Love' of the self and for those around us.

Learn more about Mimi at www.mimibland.london

'Self-image is an option, and you can change it anytime you choose'.

'As you change the inner image you hold of yourself; you will see your outer world change as well'.

Testimonials

Mimi is the definition of an exponential leader in the personal development sector.
Her latest book, *The Answer is You* will help you reach the next level of transformation if you follow her strategies.

— Lucas Gaillez Coll
Aviation Pilot

Mimi always gives us great clarity and helps to motivate our work-force with leadership, inspiration and motivation. She empowers them to achieve our business and personal growth goals.

— Mjm industrial ltd Design
build, install, maintain Quality airport and industry
infrastructure
project solutions Heathrow
xviii

You're going to love this book because it is practical, easy to read and to implement in your life. It is a pleasure to work with Mimi as she always gets results.

— Sue Griffin
Entrepreneur
Maidenhead

Acknowledgements

I acknowledge with joy and pleasure some of the great motivational teachers that have impacted my life, such as the amazing Les Brown, Dr Bruce Lipton, Oprah Winfrey, Gregg Braden, and Lisa Nichols. Thank you for giving me a new perspective on both the world and myself. You have opened my eyes to a new way of seeing both my own world and the world around me.

Also, I would like to acknowledge Victoria Beckham. She has been a personal inspiration to me for her dedication and focus on achieving her life goals. Thank you for helping me be more driven by not only being a responsible role model for all women but also being a pillar of strength through exhibiting and personifying a woman's strength.

I would also like to acknowledge the souls that are no longer here such as Mother Theresa, Martin Luther King, Dr Wayne Dyer, Louise Hay, Emmeline Pankhurst, and Marie Curie for having such a positive impact on humanity.

Lastly, I would like to acknowledge Dr Joe Dispenza for his amazing, life-transforming work towards consciously healing and awakening humanity.

To my beautiful family who surrounds me with unconditional love. x

'The two most important days in your life are the day you are born and the day you find out why'.

— Mark Twain

CHAPTER 1

Mastering your emotions to master your mind

What do I mean by mastering your mind?

I mean learning to control your way of thinking. I mean that you need to be aware of what thoughts run through your conscious and subconscious mind and then take control of those thoughts.

How do you take control? First, you need to understand the difference between conscious and subconscious thoughts and how they affect your actions, decisions and reactions.

Subconscious thoughts vs conscious thoughts

Your conscious mind is the part of your mind that is responsible for logic and reasoning. It controls your actions and intentions only when it is fully conscious. For example, when you choose to have a drink, that is an in-the-moment, conscious choice.

Your subconscious mind is responsible for all your involuntary actions. It is where all your beliefs, habits and memories are stored and where your emotions are controlled. Your subconscious mind never questions, it only obeys, depending on the information that is stored within it.

Once you understand the difference between conscious and subconscious thinking, you need to learn to identify positive and negative thoughts.

Positive thinking will always be the thoughts that uplift and motivate you. They make you feel good about yourself and where you are going with your life. For example, I am good enough, I matter, **I deserve happiness and love.** Negative thinking makes you feel bad about yourself and weighs you down in life. When your thoughts are negative, you invite negativity into your life, drawing it to you like a magnet. Most negative thoughts are fuelled by emotions; thus mastering your emotions will enable you to create a life filled with happiness, success and inner peace. It will also open you up to finding and embracing love.

Mastering your destiny by discovering your powers

You are the master of your own destiny. Life is an illusion that is created by your own consciousness; it is this concept which gives you the power over of your destiny.

What do I mean life is an illusion?

Take, for example, two people in the same situation, one person viewing it positively and the other negatively. Both have identical lives, but they each view it differently.

It is the way our consciousness perceives a situation or the sequence of situations which make up our view of life. Hence, the concept that life is an illusion of the consciousness.

If you accept the facts that life is how your mind both consciously and subconsciously perceives it to be and also that you have ultimate control of what your mind concludes, then you can be confident that you have control of what your life is.

You have terrific powers within you. Now you may have discovered these powers in the past, or you will start developing them as I guide you through this journey. Either way, those powers are constantly working on altering the illusion that we call life, even though we may not even realise that they exist.

You may wish and hope that your life will change and get better, but the only way those changes will happen is if you first create those changes within yourself. Think it, see it, and believe it; only then can you make it happen.

Once you understand how to awaken these powers, you will be able to deliberately and intentionally use them to create the happiness, success, and love that you desire. These powers will attract to you the people, opportunities, and situations that will make your life happier.

Once you have a firm understanding of who you truly are and what you are capable of, your heartfelt desires will start taking shape before your very eyes. You will be able to focus that internal power and guide it to create the life you want instead of creating the life you no longer want.

How this book will help you discover your powers

Once you work with and apply the formula in this book, you will not only have a greater understanding of how you create your life, you will also be able to consciously and deliberately direct the power within your subconscious mind to change, create, and manifest whatever you desire.

There are a few exercises in this book to help you stay focused and to begin reprogramming your subconscious mind so that you can begin to see some dramatic changes in your life.

You alone are totally responsible for where you are in your life right now. You create everything — the good experiences and the bad experiences. Own it all.

You created your relationships, your career, your successes, your finances, your failures, and your disappointments — everything that you may perceive as bad luck. You attract it to yourself, note this.

You created the positive and the negative aspects of your life, and you will create what you have yet to experience in your future. You are the master of your destiny, take responsibility for it.

You shape your destiny, whether you are doing it consciously or not. In fact, you are shaping and creating your destiny every moment of every day, right up to this moment in time, and beyond.

Everything that has and will happen to you depends on YOU.

Think it, see it, believe it, and make it happen.

YOU, and you alone, have not yet mastered how to take total control of every aspect of your life, but now you are going to begin creating your life consciously. Switch off the auto-pilot and take responsibility and full control.

In order to change and achieve the goals you have set for yourself, you have to take full responsibility for where you are right now and accept that you played a part in the way things have so far worked out for you. As the saying goes, you made your bed, and so you must lie in it.

The first key thing to accept and remind yourself of is the most important of all is this is your life, and you get to control it. Yes, you have complete control over your life.

Take a breath and say it out loud…

I HAVE COMPLETE CONTROL OVER MY LIFE!

You get to decide and choose how you will react to everything. Consciously or subconsciously, whenever you act or react in any situation, you have made the decision to act or react that way. These actions and reactions can be conditions both consciously and subconsciously.

Yes, take a moment to let that sink in . . .

Actions and reactions are decisions and decisions are choices.

Deciding to hit someone when you are angry, choosing to watch a movie or even feeling insignificant and unworthy, these are all choices that you make.

Yes! It is simply a choice. You have for a long time made choices that you have not controlled consciously and that you have not benefitted from, but now you are going to learn how to master those choices and how to use the control you have over them to make choices and decisions that are in total alignment with the happiness, success, self-confidence, and love that you want or feel that you deserve.

How do you get started?

You have to first believe that you deserve all the great things that you want, because if you don't believe you deserve those things, it will be very hard for you to convince other people that you deserve those things. You need to believe in yourself first before you can get others to believe in you.

In order to live the life that you want, you have to first learn to control your mind and to train it to receive the right information so that it can then send it to your subconscious mind to accept the happiness, success and loving relationships you want.

This means that you are not changing the picture yet, only how you perceive the picture. When your perception of the picture or scenario changes, your actions or reactions will change as well.

This means you're going to have to replace any negative behaviours that you have consciously or unconsciously turned into habits into positive behaviours and habits. All your destructive habits will need to be consciously altered. All your thought patterns associated with past emotions that keep you stuck, need to be replaced with new positive ones so that you can really live the life you choose. In order to do this, you need to decide what you want in life while releasing what you no longer want or what might hinder you on your path to what you want.

Once you start developing awareness, consciously focusing on and purposely directing your mind to think thoughts that are in total alignment and harmony with what you want in your life, you will stop focusing on all the negative aspects of your life.

As you start doing this, you will begin to eliminate anxiety, stress, and depression, which deplete your energy.

You will replace these with feelings of being in the flow and being connected to a higher power and source of energy. This will improve your creativity, increase your energy, and improve your health as well as attract the relationships you truly want in order to live a really joyful life.

Your Inner World Creates Your Outer World.

Chant it, meditate it, believe it, live by it.

Your THOUGHTS lead to your FEELINGS, which lead to your ACTIONS, which lead to your RESULTS.

Change your THOUGHTS and your BELIEFS and you will CHANGE what you are creating in your external WORLD.

Notes

'The most powerful relationship you will ever have is the relationship with yourself'.

CHAPTER 2

Beginning the process of change

To begin the process of changing your life, you first have to take a real, honest look at your life right now, without judgment. Take a close look at yourself, because you have to have a clear direction and know what you desire so that you can keep on the path to reaching your chosen destination.

You have to clearly define what it is exactly that you want to personally achieve or accomplish or experience in your life. A good way of getting total clarity is to sit down somewhere where you will not be disturbed, with a pen and a notebook, and to ask yourself these questions.

- If you could be, do or have anything in your life, what would you choose?

- If there were no limitations and you didn't have to consider anyone else and money was not an

option, what would your ideal life look like?
• Who would you be doing these things with? And
  how would this make you feel?

Once you have completed your list, you can go through
each category and choose your top ten, then you can choose
your top three in each section. That will give you a template
of what you truly desire.

Once you have clarity, you can start formulating and
planning small steps in order to begin moving in that
direction. Having a template to adhere to, will also keep you
focused on your outcome and help you to eliminate anything
that is not in accordance with your chosen outcome. Simply
acknowledge anything unwanted and move on.

Sometimes you can become comfortable living a life that
you really are not enjoying. You simply tell yourself that it is
better than nothing, or you have too many commitments to
be able to change anything.

But deep down, you know that there are aspects of your
life that are not okay. You are not completely happy, and your
mind, body, and spirit are constantly trying to communicate
this to you. Your negative emotions are signalling to you that
the life you have been living up to this moment in time is not
fulfilling you or making you feel energised and passionate
about life. You are becoming aware of a void in your life
which you have been ignoring, or you have been burying
your head in the sand so that you do not have to actually look
at your life in order to begin the process of making those

painful but necessary changes which will lead to happiness and peace of mind.

Your life can most definitely change and improve, but you have to be willing to make those changes and to take an honest inventory of your life right now.

You were born to grow and expand and to keep growing and making progress — mentally, emotionally, and spiritually. If you stop growing and moving towards what you truly desire, your life feels empty and boring. You feel like a hamster on a wheel going around and around, every day the same as the previous one, waiting and hoping that something will change, but it never does.

Nothing ever stays the same, that is the process of life. Change is inevitable, and if you look at your life, you will see just how much it has changed, even in a year. The world that we live in is always changing, no day is the same as the day before, even though it may seem as if it is.

So, in order for you to grow and move forward towards your desires, you are going to have to make some changes. Yes, some of those changes may be painful, but they can also bring you to a new level of happiness, excitement and inner fulfilment. So, begin looking at any changes needed as an exciting, new chapter of your life unfolds.

If you are afraid of making those changes because you like the comfort of the familiar, remember that everything was new at some point. Whether it was your relationships, your job, your home or your friendships — everything felt new at some point in your life, but with time, you just got used to it.

Never fear change; embrace it instead, because on the other side of fear is where your greatest potential and happiness awaits you.

You have to challenge your own limitations so that you can move beyond yourself and your old concept of how you see yourself so that you can find new things that motivate you, energise you and excite you. Imagine being passionate and eager to wake up every morning, excited about being given another twenty-four hours to design, sculpt, script, and create your own perfect movie called your life.

It's time for you to grow, expand and move beyond your comfort zone and your limitations. You begin to do that by changing what you continuously focus on and observe. As you really start to pay attention to what excites you and makes you feel passionate when you think about it, this becomes the fuel to keep you excited and energised when you wake up and begin each new day.

If you don't feel challenged in your life anymore and nothing excites you, it may be time for you to look within and make some changes. Maybe it's time to move on and change certain aspects of your life that are no longer serving you.

Change is a good thing; it's a good sign. It means that you are ready for a new challenge in your life. Expansion and growth are part of life—to give birth to new ideas and to continuously keep moving forward towards the attainment of your desires and for some new excitement in your life.

So, take some time to really reflect on what it is you really want in your life so that you no longer feel bored and frustrated but balanced, in harmony, and at peace with yourself because you are in total alignment with the life that you really want to live.

The time has come for you to stop burying your head in the sand and to start being a hundred per cent honest with yourself. You are not going to make the necessary changes in your life until you have made a firm decision to change. It will take a clear dedication and affirmation that from this day, you fully commit to changing every aspect of your life that needs to be changed so that you can finally live your life on your own terms and nobody else's.

To be really happy in life, you have to have a sense of achievement and of having accomplished something. Otherwise, you will feel unfulfilled and dissatisfied.

Notes

'Your mind is so powerful,
you can invent, create,
experience, and destroy
things just by thought alone'.

CHAPTER 3

The power of your thoughts and your emotions

It is so important that you understand the importance of and the role that beliefs play in our lives.

Beliefs are the building blocks that create our reality. They are our basic foundation, and they influence and affect our lives in ways that we have difficulty recognising.

They are constantly at work within you day and night on a subconscious level; their vibrational frequency is attracting to you the circumstances and situations that resonate with you.

All of your beliefs function on a subconscious level.

Your hopes and desires, wishes, goals and intentions are all formed on a conscious level, and you think about them with your conscious mind. But your beliefs are different; they are created at a much deeper level. They live within the subconscious mind, and every single one of us has our own belief systems. They are part of who and what we are.

There is no area of your life about which you do not have a set of beliefs and assumptions.

You have beliefs about your relationships, your earning abilities, your health, who you are and about life in general. Your faulty or limiting beliefs in any area of your life are like a faulty program on the hard drive of your computer. This program will play over and over inside of you, constantly distracting you, limiting you, and creating situations and circumstances that match the distorted beliefs that you carry around with you.

Knowing and understanding the power that these beliefs hold in your life helps you to become meticulously observant and conscious of what it is that you actually believe.

Your beliefs are formed in subtle ways

Say, for instance, your mother or father left you or died when you were a child, and you emerged from that experience with the belief that everyone you love, leaves you. By cementing that belief, you allow it to take hold of you and to create all your future experiences. You will continue to attract similar situations into your life that reinforce that belief until you change it.

From this experience, you form a belief that everyone you love leaves you, or that you are unlovable. Having this core belief results in it becoming the frequency at which you vibrate, attracting more of the same experiences to you.

So, what type of relationships do you think you will now attract to you?

Do you really think you will attract long-lasting, loving relationships to you with this belief that everyone leaves you or that you are unlovable? You will often probably sabotage your relationships, even the good ones because you are subconsciously trying to prevent yourself from feeling the pain of someone else leaving you.

But it's all an illusion; nothing has any meaning relating to a situation apart from the meaning you give it — the story you told yourself at a particular time.

You have accepted so many limited beliefs about yourself and about reality. Maybe you need to move past these false perceptions, not asking whether these beliefs are real, but asking instead:

Do your beliefs serve you or not?

Any area of your life that you are struggling with will definitely be characterised by a belief that isn't serving you.

Your beliefs are dictating and creating every situation and condition in your life.

As you change your beliefs, you can radically change your life.

Your subconscious mind is what creates your reality. It works twenty-four hours a day, seven days a week, regulating your heartbeat and controlling your breathing. Everything is done automatically because your subconscious mind essentially takes over every aspect of your life, and it does this based on your habits, your thoughts, and your beliefs.

Your subconscious mind carries a wealth of information that goes back through the entire experience of your life. It remembers every single experience that you have ever had.

Your subconscious mind creates your reality, so if you believe that it's difficult for you to be happy and to enjoy your life, your subconscious mind will create situations that make it difficult for you to be happy and to enjoy life. If you believe that it's difficult to meet people or that all the people you meet are always the same and they let you down, then your subconscious mind will make it difficult and bring you unhealthy relationships.

Whatever beliefs you hold, whatever thoughts you think; those beliefs and thoughts will become your reality because your subconscious mind acts on them. You can create the life you don't want because of one single thought that may have been planted in your subconscious mind years ago.

So why does your subconscious mind create situations that you don't want?

It's really quite simple, your subconscious mind does not know what is right or what is wrong. It doesn't evaluate whether a situation is bad for you or good for you. Your subconscious mind simply acts on the instructions it receives from your conscious mind.

Those instructions are created by your thoughts and your beliefs, the thoughts you repeated over and over and which have now become beliefs. Those beliefs are planted in your subconscious mind, and they become your reality. Everything begins with your thoughts.

It can be hard to change initially because your mind has never been told what to do. It has never received a single conscious instruction from you, and it has been able to do whatever it wants, whenever it wants to. It has basically been running on auto-pilot.

Think of your mind as a new puppy. If you never taught the puppy how to behave and never gave it instructions to follow, it would run around, causing havoc.

It is exactly the same with your mind.

Now imagine seven years have passed, and that dog has been able to do whatever it wants, whenever it wants to. Now suddenly you try to train it to behave differently. What kind of response do you think you will get? Your dog will resist any instructions you give it.

But, if you keep repeating the instructions and you keep teaching it the same thing over and over again, after a while, it will respond and act accordingly.

Your mind behaves in the same way. It's not used to taking instructions from you, but now you're going to be giving it specific, focused instructions, which you will constantly repeat. Your subconscious mind will begin to pick up the new information and start working with your conscious mind. With the new instructions and information installed, you will begin to see the changes happening in your life that you want to experience.

How long will it take for you to see the changes and results in your life? Well, that depends on you. The more persistent and dedicated you are, the sooner you will see the results you desire.

Once you develop a routine that's focused, disciplined and structured, and you follow the formula and that specific routine, you will see results. If you don't take action, nothing will change, and everything will stay the same.

You have to change your old pattern of thinking by creating a new pattern and a new way of perceiving your life. The sooner you do that, the sooner you will see changes in your life. You then have to maintain it just like you have to maintain your exercise plan if you want to stay in shape.

Remember, everything and anything is possible, you are limitless.

You may have to let go of your old ideas of what you believe you are capable of achieving, especially those that are negative and that limit you. From now on, there are no limitations, and you can be, do, or have anything your heart desires.

You have to establish a daily routine and be willing to do the exercises on a daily basis in order for you to achieve the results that you want. You get out of life exactly what you put into it.

You are now training your mind to work for you and not against you. The only way to do that is to follow the formula and work with the formula on a daily basis. This formula will now become part of your life. Eventually, you won't even have to think about it, it will just be part of your daily routine.

You will also need to take inspired action and create an action plan that will move you slowly forward to achieving your goals and desires. This will be done in small steps that will eventually build momentum and get you to your desired outcome.

You need to take full responsibility for the choices that you make and the life that you live. Only you can create the changes that you want in your life.

Once you accept that responsibility, the changes that you want will happen sooner in your life.

Your emotions

Your thoughts, emotions and feelings are what create your life experiences. So, in order for you to begin to create new life experiences, it is essential that you understand the importance of your emotional guidance system.

You are controlled by your emotions. If you consciously choose the emotions you want to take into your future experiences, you can, with awareness, maintain a positive state of being, guiding you towards the life experiences you want. Most people are living from the emotions of their past.

Those past memorised emotions — fear, anger, jealousy, guilt or regret — will continue to create their future experiences resulting in similar outcomes to those they experienced in the past.

You have an incredible ability to remember the past and bring up memories, visions, and pictures in your brain, which cause you to relive those past experiences as if they were happening right now. Every time you do that, if those emotions are highly charged, they can have a very negative effect on your life and on your body.

Past experiences have caused you to develop specific emotional patterns that are responsible for your emotional outbursts and negative responses as well as for creating your health issues.

That is why it is imperative that you do not hold on to any negative emotions from the past or try to bury or ignore them. They will continue to have an effect on your life, even if on a conscious level, you may think they are not, believe me, they most definitely are.

Your emotions are simply there to guide you, inform you and reflect back to you what you are going to attract and experience in your life. They play a crucial role in the world you live in and in your day to day interactions with other people.

Your emotions are simply energy. They are just sensations in your body that are constantly trying to communicate with you and give you feedback to help you move towards living, creating, and sustaining a beautiful, happy life.

Understanding your mind

Your mind is designed to do three things. It regulates and keeps your body alive; it registers and catalogues everything that has happened to you so you can have memories; it's designed to always protect you.

The way your mind protects you is by magnifying risks and by magnifying every conversation and situation.

Your subconscious mind is like a huge memory bank. Its capacity is virtually limitless, and it permanently stores everything that happens to you – all your feelings, thoughts, behaviours, and memories that are outside of your conscious awareness.

Most of the memories stored in your subconscious mind are of unpleasant experiences associated with pain, anxiety, conflict, and self-loathing. The function of your subconscious mind is to store all past experiences and to retrieve the information when required. Its only job is to constantly ensure that you respond to situations and circumstances in exactly the way you have been programmed to.

Your self-image and concept of how you see yourself will remain consistent with everything you say and do. It literally controls every aspect of your life and the meaning you give to everything. Your subconscious mind is powerful, and it does not reason or think independently. It just obeys every command it receives from your conscious mind.

Your conscious mind is your thinking mind. It is where all your wishes, dreams, and desires are formed. Your subconscious mind is the controller, and it doesn't matter how much new information you receive. It is programmed to force you to 'see' what it wants you to see and it gets its information and points of reference for every new situation in your life from all your past memories and experiences.

Memories of negative experiences that have not been updated with new information will not let you think or perceive in any other way than you always have done.

'The subconscious is not selective; it is impersonal and no respecter of persons. The subconscious is not concerned with the truth or falsity of your feeling; it always accepts as true that which you feel to be true. Because of this quality of the subconscious, there is nothing impossible for man. Whatever the mind of man can conceive and feel as true, the subconscious can and must objectify. Your feelings create the pattern from which your world is fashioned, and a change of feeling is a change of pattern'.

— Neville Goddard

'The definition of insanity is doing the same thing over and over again, but expecting different results'.

— Albert Einstein

Notes

'What you think, you become.
What you feel, you attract.
What you imagine, you create'.

— Buddha

CHAPTER 4

How you create your reality

Your life is a perfect reflection of you. If you do not like what is happening in your outer life, then go within and make the changes.

You are the sole creator of everything going on in your life — all your thoughts, emotions, fears, doubts, and worries. All your heart's desires, your reactions, and actions all come from you.

You have created a vibration of energy and life is simply and always responding to you according to your vibration of energy.

'You create everything that happens in your life'.

— **Mimi Bland**

Every single thing is energy.

It is a basic, universal truth that everything is energy. Your body, your bank account, your job or business, your relationships and your emotional state, the house and life you desire as well as your health — all is simply energy.

It is all just a vibrational frequency. Energy cannot be created or destroyed. Every time you say you want to create more money or create a house or create happiness, technically, it's not the right way to say it because energy isn't created. What you want already exists. It is just on a different vibrational frequency to the one you are currently on.

Nikola Tesla said if you want to understand the universe, you have to think in terms of vibration and frequency.

The first step is to think in frequency, what you want already exists.

Your body is just one per cent of who you are. All of reality is just an energy field, and everything is just a field of vibration. Your frequency pattern creates and holds the people you attract into your life, your financial status and all the situations and circumstances that come into your life.

Every single person has a frequency pattern. Just like a radio station transmitting a signal at a certain frequency, if you are not tuned to that particular frequency, you will not be able to hear the broadcast from that particular station. Your mind works in the same way. You have to be on the same frequency, the same vibration, in order to attract what you desire into your life.

Look at your body and start imagining it as an energy field. So, what determines your energy field? Well, it is your intentions, your perceptions, and your consciousness.

Your body is made up of trillions of cells that are made up of energy, and they respond to your consciousness, your perceptions, how you see everything, and the meaning that you choose to give to every situation and circumstance in your life.

Your assumptions create your energy field, which extends out beyond your body and creates your vibration. That is why you can pick up a good vibration or a bad vibration, because everybody is just an energy field. If everything already exists, you just need to match and stay in the frequency of what you want, and you can't help but get it.

You have to begin by suspending your disbelief and by making a conscious choice to set an intention of what you desire to have in your life, regardless of what that might be.

Remember, there are no limitations, only the ones that you set in your own mind. You literally can create anything if you truly believe you can.

You have to form a clear picture of what you want in your mind whether you want a new job, a successful business, a new home, a particular relationship, to heal your body, or to attract more money into your life.

Get total clarity and clearly define what you want because you are putting that frequency pattern into your energy field, and you are moving from a lower vibration to a higher one. Remember, the higher vibration is a frequency pattern that already exists. What keeps you separate from receiving what you want is your beliefs. If you truly believed a hundred per cent what you wanted, without any resistance or negative thoughts, you would be able to manifest your chosen desires instantly.

But, if you believe that you are separate from this energy and that it takes hard work and years to get what you want, you're solidifying a frequency of limitation. You can only attract things that match that pattern. That is why it is so important to suspend your disbelief.

What happens when you suspend your disbelief? You will begin to match the frequencies that you want. The moment your beliefs match the higher frequencies of the job, the money, and the new opportunities, the activation and projection of that higher frequency will attract the conditions, circumstances, and people to your life that match the new signal that you are emitting.

This new state of conscious awareness becomes your new point of attraction from which you experience the world.

But, if you are constantly observing your current reality and thinking, you can't do it, it's going to take years, and you don't think it's going to happen, then... it won't. The more you focus on the absence of what you want, the more you will think about past situations in which you didn't get what you had hoped for.

By focusing on the past and recalling negative memories, you remain in a place of not being in sync with what you want. The frequency of your desires is much higher, and you are vibrating much lower. You are in a state of misalignment and resistance. Everything is coming in response to your consciousness and perceptions.

You already have what you want. If energy is not created or destroyed, that means that everything already exists. You just cannot see it yet with your five senses in this three-dimensional reality.

You already have the job, the money, the relationship, the health, the life you desire. If that were true, what would you be thinking?

How would you feel?

Think about it, all your desires have been fulfilled, your life has completely changed, you have received a lump sum of money, you have met your soul mate, and he or she has told you that they love you. You have bought your dream home or dream car, and your body has completely healed.

What would you be saying to yourself and to other people?

'OMG! You're not going to believe what has just happened to me! I'm so excited! I woke up this morning and I received an unexpected cheque in the post. It has literally transformed my life'!

OR...

'I was talking to someone I know in the supermarket today, and they are friends with the managing director of a large company that I have been trying to get in touch with for months without success. Well, he made a call right there and got me an appointment to see the guy, and guess what? I got the job! It's amazing! My dream job, my present salary quadrupled! I am so excited'!

Creating sentences as if what you desire has already happened is an excellent exercise to practise bringing the reality you want into the now. By doing this, you take your desires from the future and embody them in your present consciousness.

Then, practise gratitude and simply let go. Surrender your desires to a higher frequency and a higher level of consciousness.

Remember, 'Everything is frequency and already exists'. This is where you need to develop faith because your habitual mind will start looking for signs to validate that something is happening, things are changing.

But, if you keep looking for signs and nothing seems to be happening for you to observe, you may begin to lose hope and start speaking negatively about what you want.

'I didn't get the house.'

'I didn't get the money'.

'The man or women I was hoping to meet never showed up'.

Now, you have gone back to a lower vibration, as your consciousness is observing what is not happening the way you want it to, and you are now trying to force the situation.

Quantum physics is explained in the Bible:

'To those that have, more will be given,
and to those that do not have, even that which
they have will be taken away'.
        — **Matthew 25:29**

So, the more you are panicking and worrying about what is not happening, the more that which you have will be taken away.

You are literally pushing the very things you want to have away from you through your constant attention to the absence or lack of what you desire and your constant attention to your present reality and what you are currently perceiving with your five senses. Your outside world is just a holographic representation; it is a mirror that reflects back to you, your present state of consciousness.

So, those that have receive more because they are constantly in a place of total gratitude. They vibrate at a high frequency, whether consciously or not. They understand that everything is simply a frequency and that their gratitude and beliefs keep them vibrating at a high frequency.

When you think of yourself as a spiritual being in a physical body that has an energy field based on your vibrational pattern, frequency, and the feelings you are emitting out into the quantum field, you will begin to understand how you attract the circumstances, situations, and people into your life that match your energy and your frequency.

There is a place, it is a frequency, which you are tuning into with your own imagination. The moment your belief matches that state of being, that vibration, and frequency, you will become a complete match to that energy, and you will draw what you want, like a magnet, into manifestation.

Start to practise saying thank you and know it is already here. The conditions and circumstances will arrange themselves to attract that which you want to manifest into your life, and you will begin to witness just how powerful you really are.

It all comes down to your imagination, as you hold the image in your mind, what you are doing is, you are starting to change your energy field and your frequency pattern.

When you can practise gratitude and excitement, knowing it's already here, your frequency pattern matches and fuses with that reality and attracts all the synchronicities into your life.

Notes

'Self-image is an option, and you can change it anytime you choose.'

'As you change the inner image you hold of yourself, you will see your outer world change as well'.

CHAPTER 5

Reconstructing your self-image

You have 24 hours every day to recreate your life, a life that you choose and want to live.

Every single day you are given a new blank canvas, another 24 hours to design, create, and live your dream life. You have to decide that feeling good is now the most important requirement of your life.

You have to change your mindset and take control of your emotions so that you can achieve your ultimate goals. Remember everything is coming from you and your thoughts. As you begin to change your thoughts, your life will begin to change.

Your brain is just a processor of reality. Everything you perceive with your five senses and that you give meaning to as a result of your past experiences, all your stored information — that is the reality and life experiences you create.

That is why it is vital to keep gathering new information that will break down the foundations of your beliefs, perceptions, and past memorised emotions. This will enable you to transcend your programmed limitations and to operate from a whole new level of mind.

Whatever you keep focusing on in your life is your emotional set point. Feeling that you do not have something, or lack a certain aspect in some way, and thinking those thoughts over and over keep you identifying with them. Now it's time to begin to create a whole new vision of yourself and the life you want and consciously choose to live.

Remember, where you are right now in your life is only a vibrational interpretation of how you have been using and directing your energy, where you have been placing the majority of your focus, and how you have been feeling about every aspect of your life.

So, go back to your notes on exactly what you want in your life. You are now going to create a new vision, a new template so that you can create a new vibrational frequency that matches what you want so that you can begin drawing those experiences into your life.

Your thoughts are the electrical charge, and your feelings are the magnetic charge. How you think and how you feel, broadcast an electrical, magnetic frequency that influences every single atom in your life.

Your thoughts send the signal out, and your feelings draw the event back. As you change your energy, your life will change.

Your self-image

First, you have to change the image of how you see yourself because your self-image is the key to living your life without any limitations. Your self-image is simply your perception of who you think you are. Up to this point, it has been created by you from the beliefs you accepted from other people, the times you felt humiliated or fearful, from your failures and successes, when you felt insignificant and unworthy, and how other people reacted to you, especially when you were a child. Through all those experiences, you mentally created a picture of yourself, and you inject those beliefs about yourself into that picture. It then becomes your truth, and you never question whether it is right or wrong. You just act as if it were true, and it controls every aspect of your life.

It controls what you think you can or can't accomplish, how other people respond to you and treat you, how much money you will allow yourself to make, all your actions, and how you feel. All of your abilities and behaviours are always in total harmony with how you feel about, and see yourself.

So, you will act like the person you perceive yourself to be, regardless of how much you consciously want to change. You will default back to your programmed, self-image, and you will look for and find circumstances to prove to yourself that how you saw yourself was right.

If you believe that you have a problem learning new things or attaining what you desire-regardless of whether it is more money, better health, a loving relationship, or happiness, it is because it is not in accordance with how you see yourself.

So, reconstructing your self-image is the key as you cannot outperform your own concept of self.

Reconstructing, changing and expanding your self-image so that it is in alignment with your new vision of your life, regardless of your age, involves changing your thoughts, your concepts, and the image of how you see yourself. As you make these changes, your circumstances will also change.

Whatever target you set yourself to achieve, you can achieve. It starts with the habits you form and seeing yourself as a person who can have and be anything you set your mind to — wealthy, in a loving relationship, with your ideal body, and perfect health.

You start acting like someone who can easily accumulate wealth, who already feels whole, who feels adored, and is in love with life. Now your energy changes and you move into a higher vibrational frequency. Your self-image begins to change, and you start attracting more opportunities to you that match your new vibrational signature because everything you want will come to you in proportion to how you see yourself. Now you will attract according to the new self-image you have of someone who can have everything.

'You are someone who easily gets what they want'.

Everything is a mindset, a consciousness shift. It's all simply a change in perspective and how you identify yourself with your chosen goals.

Real, lasting transformation occurs through self-realisation and conscious awareness.

Processes to reprogramme your subconscious mind

Exercise 1

Your eyes are said to be the windows of your soul, revealing your innermost thoughts.

Never underestimate the power of mirror work. Looking directly into your eyes in a mirror releases a tremendous power.

I personally healed my body and completely renewed my mind through the practice of mirror work and consciously developing self-awareness by seeing my true self in a mirror.

Stand in front of a mirror. If it is a full-length mirror, great, but if it isn't, it will still be effective.

Stand up straight with your shoulders back, suck in your stomach, and hold your head up. Now take three or four deep breaths until you feel a sense of power, inner strength, and determination.

Look deep into your eyes, into the very depths of your soul, and tell yourself that you are going to get what you want. Say it out loud so that you can hear the words you are speaking.

Practise mornings and evenings, and you will be amazed at the results.

As you stand in front of the mirror, keep repeating positive affirmations and statements.

Examples:

> 'I am an outstanding success. I totally love my life, and for this and so much more, I give thanks'.

> 'My body has completely healed and restored itself, and for this I give thanks'.

Remember that every idea presented to the subconscious mind is going to take shape and manifest in your external life. The more powerful your words are, mixed with real heartfelt emotions, the more you will see real magic happening in your life.

You can use the mirror technique for any subject. You can even reconstruct your whole self-image and become any person you choose to be.

Exercise 2

Pick one personal character trait that you wish you had, something that you believe you lack. Maybe you wish you were confident or motivated or determined or happy. Spend five minutes every day visualising yourself having all of those qualities and all the things that would be happening to you if you were confident, motivated, determined, and happy.

Lose yourself in that image and really feel what it would feel like to be that person.

Spend five minutes every day visualising yourself feeling confident, happy, motivated and determined. See yourself in scenarios where you are now that person. See how you are walking and talking to people, see yourself smiling and shaking people's hands. Keep playing out these new scenarios so that they become embedded in you, becoming your new personality traits.

Exercise 3

Get an index card or a large post-it or fold a piece of paper to a small enough size to carry with you and write these words on it.

'I am so happy and grateful that (fill in your desire)

Examples:

'I am so happy and grateful that large sums of money now come to me easily'.

'I am so happy and grateful that my life is now full of love'.

'I am so happy and grateful that every cell in my body now vibrates with energy and perfect health... thank you, thank you, thank you'.

'I am so happy and grateful that I now love myself, exactly as I am'.

Exercise 4

Consciously implanting a new way of thinking. Ask yourself daily:

'What thoughts do I want to put my energy behind'?

'What do I want to believe about myself'?

'What would it feel like to be really happy'?

'What new experiences will I experience in my new life'?

'Who am I going to rehearse becoming every day'?

'What if a miracle happened in my life today! What would I see'?

Why is this important to do? Because when you are a person who doesn't have what you want, whether it is money, health, relationships, or success, you identify with that reality. You have about 65,000 thoughts every day, and 95 per cent of those thoughts are subconscious and constantly affirming the same results in your life. Most of them are unconscious, but your old thought patterns and beliefs sabotage what you are trying to achieve in your life today.

That is why you are now going to create a whole new belief system. Your old thoughts are stuck on what is happening in your life right now, your present reality of what you don't have. Your constant focus on what you don't like or want is blocking you from allowing what you do want to flow into your life.

All these tools are going to begin to embed a new belief system within you:

'I easily get what I want'.

'Money flows to me easily'.

'My life is full of love'.

Carry that card with you everywhere you go to keep reinforcing the new ideas inside of you. Repetition is the key until it becomes natural for you to think that way.

You have to start re-suggesting new thought patterns to build new habits, new thought patterns that will change the way you feel about yourself and how you see yourself.

You have to begin to reprogramme yourself until it becomes ingrained in you, that the old you that was programmed with faulty information from external sources and has now been replaced with a new awareness of just what you are capable of achieving. Your new, empowered self-image, new perspectives, and unshakable, rock-solid belief system will enable you to become limitless and to achieve whatever you truly desire.

The shifts will happen quickly once you begin to reprogramme yourself to live the life that you want, and you begin to see yourself as the person who has already attained it.

Not the life other people want you to live. Not the life you are living at present, but the life you choose and want to live.

Notes

'Everyone thinks of changing the world, but no one thinks of changing himself'.

— Leo Tolstoy

CHAPTER 6

Creating and sustaining a powerful vibrational signature

Your unique vibrational signature always feeds back to you what you are constantly seeding within yourself through your chronic thoughts, your conversations, your memorised emotions, beliefs, and perceptions.

Now that you have begun to reprogramme your subconscious mind, you will start seeing positive changes in your life.

To make sure you don't fall into your old patterns of negative self-talk that usually result in you sabotaging yourself and creating inconsistent results in your life, I have added some more techniques that will enable you to stay on track so that you can enjoy a lifetime of positive results.

1. Focused attention

Focused attention simply means being able to consciously direct your attention, becoming aware if and when your mind begins to wander, and then being able to redirect your attention and focus back towards what you desire.

The subconscious mind accepts what your conscious mind consistently focuses on, so it is important to focus only on the positive aspects of yourself and your life. Your subconscious mind will continue to accept and affirm these new thoughts as your new truths.

Remember, your subconscious mind does not consider whether your thoughts are good or bad, true or false. It responds mainly to what your conscious mind focuses on and what you feel when you focus on a specific mental image.

2. Stop overthinking

When you feel stuck, it is because you have developed the habit of thinking too much. When you keep excessively talking about and analysing things in your mind, you will end up confused or doubtful.

When you spend too much time analysing and worrying about the problems you have, you have less time to focus on the solutions. Instead of taking action, you feel constantly stuck in your thoughts that are not in alignment with what you want.

The more you keep thinking and focusing on the same thoughts, the more you will fill your mind with worries, doubts and fears that will negatively attract the opposite of what you actually want.

You cannot solve the same problems, with the same level of mind that created them.

Get off the subject that is bothering you for a while and go for a walk or fill your mind with something that uplifts you. Your mind will then be in a more receptive state to look at the problem from a different perspective and to find new solutions to solve the same problem.

3. How to raise your vibration

You have total control over your vibration, and your vibration is always being shown to you by the emotions that you feel. Your vibration is the result of the thoughts you think, and the thoughts you think are always about what you are placing your focus on.

You are always offering a vibration, which is always being projected from within you with every thought and feeling that you think and feel.

You are always in control of your thoughts, and you can deactivate any thought, anytime you choose. The way you deactivate those thoughts is by activating other thoughts.

You can at any time take a thought that is active, but that does not have a lot of emotion attached to it, and within seconds you can replace it with its exact opposite.

You cannot change a really active, negative thought to its complete opposite once that negative thought has built up momentum. You can begin to slow down your negative thoughts by withdrawing your attention from them, or by approaching the problem in a more general way, avoiding the specifics of it.

4. Stop talking negatively

Be careful of what you speak into your life, your words have tremendous power.

Your words matter. Your words, when mixed with your dominant beliefs, will become your reality. Every time you speak negatively about yourself, that belief grows stronger and soon, it will become who you are. Stop talking about the things you don't want to keep experiencing in your life. You are just keeping the momentum going, and that negative vibration attracts more of the same back into your life.

Make a decision to talk more about the things you do want in your life.

When you start observing yourself talking about the thing you don't want, you can simply stop mid-sentence and redirect your conversation.

You will immediately raise your vibration as you replace those negative conversations with positive words instead.

Start speaking your dreams into existence and believe and trust that they are now on their way to manifesting in your life experience.

5. Every day set an intention to only focus on things that make you feel good

Intention-setting can become part of your life-changing practices. Becoming intentional is to become 'on purpose'. By setting intentions, you start to become more aware of your thoughts and your behaviour patterns each day.

Eventually, the pathways of your brain shift, and throughout your day, you will see your intentions manifest. The more energy you put into your intentions, the more aware you will become.

Becoming intentional is an important part of creating changes in your life. Setting intentions is about developing an awareness of the things you wish to experience.

'Today I want to be more focused at work'.

'Today I will only focus on things that make me happy'.

These are examples of intentions.

Set a new intention every day and throughout your day, think about your intentions. It will take your mind a while to adjust, but over time you will build momentum, and you will start seeing your intentions manifest more quickly.

6. Gratitude

Gratitude is the most powerful of all human emotions. Being grateful can change your life because it makes you appreciate what you have rather than what you don't have.

Gratitude is the most powerful source of inspiration that you can apply daily to your life. It is the creation of love, joy and happiness.

Practising daily gratitude and reflecting upon the things you already have in your life and that you are thankful for. This helps you to experience more positive emotions, feel happier, and to express more compassion and kindness.

Gratitude connects you to your heart, not just your mind, and that is where real change begins. Your heart is like a giant magnet. It is the key to creating what you want and releasing what you no longer want. Your heart is the magnet, not your mind.

You have to connect your mind and emotions so that they are in complete harmony with each other. When that happens, miracles happen.

7. Gratitude journaling

Write down five things, every single day, that you are now grateful for.

Set your preferred time. I prefer first thing in the morning before I start my day, as it sets my vibrational frequency to attract more things that I am grateful for into my life. You will begin to create a new habit of automatically writing in your gratitude journal every single day.

8. Meditation

I cannot overemphasise the importance of daily meditation. The practise of daily meditation is to begin to 'know thyself'. It helps to slow down your analytical mind and connects you to a higher power. It bypasses your ego and your identity and connects your breath to your heart.

Meditation enables you to release all resistant thoughts, and once those resistant thoughts are gone, you automatically move into a state of appreciation.

Your heart is the most powerful transmitter; it is far more powerful than your mind. As you learn to go within, you will increase your capacity to create and to stay present and in the now.

Close your eyes and breathe in deeply through your nose. Suspend your breath for a few minutes, hold your breath, and then release it out through your mouth slowly.

As you begin to slow down your analytical mind, you will begin the process of forgetting about yourself, forgetting about your environment, and forgetting about time. You become completely present and at one with yourself and with source energy. You will begin, after practising for a while, to connect to a greater, unseen power, a greater intelligence that is all-knowing, all-loving, and forever present in each and every one of us.

Meditation allows you to place your undivided attention on what you want to experience and manifest in your life. It completely centres and strengthens you internally. In that peaceful, loving state, you get to clearly define exactly what you want. Once you have visualised it, felt it, and totally immersed yourself in the experience, becoming one with it, you simply let it go and give thanks to the universe, knowing with complete faith that it is on its way.

Always finish your meditation with heartfelt gratitude and thanks.

Thank you, thank you, thank you.

9. Letting go

Let go of having to control every situation and the need for you to always have things be a certain way in order for you to feel good.

Let go of all the situations and people that you cannot change and instead only focus on the things that you personally have the ability to change.

Make living in peace more important to you than having to be right.

Conclusion

Now you are developing the art of awareness and observation. You will begin to notice that your life is changing. Certain things are not coming up as much in your experience because you are beginning to direct your thoughts around situations and events.

You have started directing the momentum of your vibration, and you are now beginning to have fewer, active thoughts that are in opposition to what you want. You are now beginning to focus and are not so distracted by everything that is going on around you.

Tune into your emotions, on how you are consistently feeling on a daily basis, and let them guide you towards the attainment of your desires.

When you are feeling negative emotions, it is an indication that you are not in harmony with what you truly want. Something is important here, or you would not keep feeling this emotion. Your feelings are simply trying to communicate with you.

Ask yourself:

'What is it that I want'?

Your answer will be:

'I want to feel'; or,

'I want to be'; or,

'I want to have'.

These questions will stop you living in the past and will bring you back to being present and in the now so that you can create new emotions and thoughts.

Now that your vibration is higher, you will begin to experience new ideas that will begin to flow into your mind. As you begin to withdraw from the chaos of your own mind, you will find a feeling of peacefulness and wellbeing within yourself.

You will now have the ability to deactivate lower vibrations and to consciously and deliberately activate higher vibrations instead. You will realise that you are the creator of your reality and the manager of your own vibrational offering, which is attracting and creating your entire life. The universe is always responding to your vibration.

As you quieten your mind more often, you will choose to think of things that soothe you and make feeling good your number one priority.

Question the importance and validation of your negative thoughts:

'Do those thoughts have any value for you'?

'What will you benefit from continuing to think about those negative situations'?

'Do those thoughts matter in the grand scheme of things'?

'Will those thoughts really matter in one week, one month or one year'?

Your answers will help you to simply let go of those thoughts you keep thinking.

Some old thought patterns may occasionally reoccur in your life. When this happens, just redirect your attention to anything that makes you feel good. Practise letting go and replacing those negative thoughts with thoughts that make you feel good. Eventually, those negative thoughts and beliefs will begin to dissipate and disappear from your life.

Notes

'I AM'

'These are two of the most powerful words, for what you put after them shapes your reality'.

'"I AM" is the name of God'.

CHAPTER 7

The Answer is You!

You were born a powerful creator able to choose what you want to experience in your lifetime and to reclaim your personal freedom. You must learn to pull back from your life to enable you to regain your personal power. Then you can re-emerge in control of yourself, your mindset and your emotions, no longer reacting to external situations and conditions.

You will have a firm understanding of the importance of your emotions and the significant role they play in offering you guidance so that you can redirect them, change, and modify them so that they are in harmony and alignment with your desires.

You have the ability to be steadfast and secure in your mindset, your intentions and your inner power so that you can be the powerful creator you were born to be.

Your consciousness, which is being aware of and responsive to your surroundings through your thoughts, your sensations, and your feelings, as well as the meaning you personally give to everything based on all the knowledge and information stored in your mind from all your past experiences, will always attract to you exactly how you feel about yourself because your consciousness creates your whole reality.

Why do you have negative experiences and feel feelings of pain and suffering? Because, if your life was perfect, you would have nothing to expand towards and to have opportunities to grow and experience life. That is why you came here in the first place. To experience all aspects of life.

If you hadn't experienced painful relationships or if you never suffered from financial hardship or developed an illness or disease, you wouldn't have had the desire to experience the opposite of those situations and look for new ways to solve your issues and new ways for you to feel happier about yourself and the life you want to experience.

You consciously decide and choose to release the feelings of insecurity and unworthiness, the fear of abandonment and the negative outcomes. Instead, you learn to trust in and understand the process of life. You realise the importance of disconnecting from your life and going within to reconnect with your eternal life force which is ever-present and all around you, waiting for you to resurrect and give it life so that it can heal your mind, body, and every aspect of your life.

You were born to come here to create with the eternal power, knowledge, and wisdom of awakening and remembering who you are and connecting to the powerful life force that is within you.

Once you accept total responsibility for your life and for this great power within you and all around you, you will understand that it is just reflecting back to you exactly what you are constantly, vibrationally, broadcasting via your thoughts and emotions.

The universe always gives you back how you feel about yourself, so... how do you feel about yourself? What are you calling into your life?

When you lose your connection to your power, you feel powerless, weak, and vulnerable. You look to other people and material things to make you feel something, to help you to feel loved, validated, worthy, and secure.

The only way you will truly be happy is by freeing yourself from needing external situations to feel anything.

As you connect to your power and learn to connect to it at will, you will set yourself free and create a powerful, conscious state of being.

Your power is connected to the amount of life force you express, and it expresses itself through your body, your emotions, and your thought patterns as well as through the way in which you perceive yourself and deal with your life.

You are here to awaken to yourself and your own empowerment. Your empowerment is the life force within you. That life force is simply... 'Love' — love for yourself and all things as well as compassion for the trials and errors of mankind.

As you now learn to look at the world from a detached state of being and feel compassion but are detached emotionally, you now observe other people's lack of understanding and wisdom. You become free and reclaim your own power as you begin to detach from reality, rather than being sucked into what is happening around you.

As you accept that your consciousness creates your reality, you can alter your destiny.

You have experiences in your life to show you what they feel like, but it's what you do with those experiences and feelings that matters.

What new experiences did those experiences make you create?

What beliefs did you establish?

How did you perceive yourself and the world you live in?

Every single experience has a lesson within it to help you grow and learn about yourself.

You can visualise, affirm and desire, but your life will not change greatly, because life is constantly responding to how you feel about yourself.

Never compare yourself to other people or change to make other people like you.

As you keep letting go of being in control, and as you let go of fear and manipulating other people, you will step into your authentic self with clarity, intuition, and a heightened awareness of who you are. You will have new perspectives on your consciousness and your reality.

Love, appreciate, and accept yourself. Always stay true to who you really are and what living an amazing life means to you. No one has any power or control over your life. They only have the power and control you have allowed them to have.

Put power back into your body and make it strong. Unhealthy foods and negative emotions over time weaken and destroy your power.

Let go of having to constantly have an opinion. Let go of control and limit negative, restrictive situations and people who leave you feeling drained, and love your immune system and energy.

You are now creating a new model of reality that you have chosen to live in, so choose things that are exciting and make you feel good. Use your consciousness as a laser beam of intention. What you place your focus and your attention on will become part of your life experience.

You release a power within you when you are happy and make peace with yourself. Your journey is unique, and you have the opportunity every day to come out of any position you find yourself in.

You are here to experience new experiences so that you can grow and expand and become your true powerful self. You are not here to experience pain and suffering or any type of control or manipulation. You are not here to obey rules and regulations or to have to comply in any way, but to let go of all fear and doubt about what might happen. Stay connected to your life force, totally safe, aligned with your power and the power of the universe, so that you can create and shape your life to reflect what it is that makes you happy and brings joy into your life.

When you remain in situations that make you miserable, depressed, and feeling empty and incomplete, you feel stuck and you diminish your own happiness and joyful life experiences.

Your choices may not make other people happy, but you are not here to comply with their wishes regarding how you should or should not live your life. You have to be true to yourself regardless of your circumstances or situations. You are and always have been a hundred per cent free to think and change your perspective on any subject in your life, free to make new choices and decisions, and free to expand beyond your own self-imposed limitations.

Loving yourself is powerful, not in an egotistical way, but by being non-judgemental and non-critical, and by just accepting yourself for who you are and where you are in your life.

Develop your power and your perceptions so that you are really aware so that regardless of who you are around, you can look at other people and accept them for who they are, never trying to dictate to them or forcing them to adhere to your way of thinking or to control them so that you can feel good about yourself.

Always come from a place of absolute love and compassion. Learn to let go and release past hurts and resentments because you will block your own life force and energy flowing to you. Take an inventory of your life and examine what is really important to you. Let go of what is no longer serving you, and you will begin to move into a higher power and a higher vibration, which will allow new people and opportunities into your life.

Your only task in life is to live and create a life filled with love and happiness.

Make that your daily intention, that whatever you do today, you find love and happiness.

Once you set that daily intention, you will be drawn to the right opportunities as the Law of Attraction brings together all the necessary components needed for your desire to manifest into your reality.

As you begin to look for positivity in your life, that is what you will begin to experience more of.

Words do not teach, only your life experiences do. That is why you must try it for yourself. Once you begin to coordinate what you are doing mentally, how you are feeling, and what your life experiences are reflecting back to you, you will understand and apply the formula to realign your thoughts and feelings, redirecting them towards creating what you do want to experience in your life.

Imagine for a moment… What if you woke up tomorrow morning and ninety per cent of all the rubbish in your head wasn't there anymore? What if the majority of your negative thoughts, emotions, feelings, judgments, and destructive, limiting, sabotaging habits and patterns were totally eradicated from your life?

What would your life look like?

What if you truly knew that there was nothing wrong with you, that you are not broken or damaged or limited in any way, and that you could be, do, or have anything you truly desired?

Ask yourself whether you are willing to allow something different and new to emerge in your life.

Right now it's time for you to allow your greatness to show up and for you to boldly and confidently state that you will no longer accept any limitations or judgments from other people, that you will now reclaim your life and live it according to your highest values and according to what's truly important to you.

You will stop seeking approval from other people, and you will become the master of your own destiny.

Today is your day and now is your time to step forward and start taking small steps in the direction of fulfilling your dreams and aspirations by eliminating any negative feedback, while holding on to your vision, knowing that as long as you keep it alive and it's in total alignment with your highest values, it has to eventually become part of your life.

Your only work is to stay congruent and to know yourself, to love yourself, and to live your life on your terms and in your way.

TODAY is the day! ...and NOW is the time for YOU to finally set yourself free, master yourself, take control of your emotions and step into your creativity and to take back your personal power.

There will be no more pain, no more judgment, no more fear, no more feelings of unworthiness, and no more doubting yourself and sabotaging your own happiness. No matter what the situation is, in this moment, you can find a way to change it.

You have the power and the ability to create what you want and to change, expel, delete, and let go of what no longer serves you.

You ARE a powerful creator. Do not waste one more second of your life doubting yourself.

The essence of our lives is to create a life that we love and to live that life in total gratitude every single day. We are born to live an inspired life, to follow our own spiritual path, and to awaken to our own unique gifts so that we can share them with other people.

Nothing in this world is more important for your survival, your physical and mental wellbeing, and every aspect of your life than the ability to Love and to live according to what you value most in life.

Love is the key; gratitude, appreciation, and a joy of life itself is the formula that will heal and transform and change every aspect of your life.

Give yourself permission to live your life according to what you truly value most in life. Stay open and receptive to all avenues that present themselves to you.

Start saying YES to your greatness, YES to your success, YES to your happiness, and YES to living a full and happy life filled with an abundance of Peace, Love. and Happiness.

The answer you are really seeking is... 'YOU'!

Sending you all an abundance of peace, love, and happiness, Mimi x

Notes

Believe

in yourself

&

you will be

Unstoppable

Questionnaire

Self-confidence, Self-Image, Self-Esteem, Self-Acceptance

Self-confidence

You cannot outdo your self-image — the way that you see yourself — it is what you believe about the results that you are able to create in actual situations. It's the expectation of adapting to whatever circumstances you are in and engineering positive results. The person with unshakable self-confidence believes he can turn horse manure into gold.

Self-image

Self-image is the picture you hold of yourself in your subconscious mind, and it determines your areas of possibility as you cannot outdo your self-image.

Self-esteem

Self-esteem is how you feel about yourself. Specifically, what your capable of achieving and what you deserve to achieve.

Self-acceptance

Self-acceptance is freeing yourself of inhuman expectations of perfection, such as comparing yourself to others and coming up short. It is working to develop your best skills and abilities and not to compete with anyone else.

Set yourself up to win

Honestly access where you are in your life and who you are. Appreciate your strengths and your weaknesses and constantly work to improve them.

Building a positive self-image

Accept yourself where you are now and tolerate any imperfections in yourself.

Creating a new self-image.

Do not punish yourself or beat yourself up for any past mistakes you have made.

Goal-Setting Exercise – Defining Your Dream

Answer the following questions as truthfully and in-depth as possible.

1) What do you love doing so much that you would do it for free?
2) What are you truly passionate about?
3) What can you do better than anyone you know?
4) If you were a multi-millionaire, what would you do with your free time?
5) What did you want to do when you were a child?
6) What would you do if you were absolutely guaranteed success?
7) How much money do you want to earn in your lifetime?
8) What skills would you love to master?
9) What do you want to give back to the world?

If you could be, do, or have anything – what would it be?

Write down EXACTLY what you desire.

Transformational Affirmations

- My future is in my hands.
- I take advantage of all the opportunities surrounding me.
- My potential is limitless.
- Better things are always coming to me.
- Love and blessings are chasing me down.
- I can do anything I set my mind to.
- Nothing is impossible for me.
- I love myself enough to live my best life.
- My ideal life is waiting for me.
- I always pursue my ideas.
- I work on my goals every day.
- I move forward in spite of setbacks.
- I am consistent, determined, and focused.
- I attract positive things and positive people.
- I choose to do great things today.
- Every day, my life is getting better and better.
- Wonderful things are happening to me today.
- I am so happy and grateful that great things are heading my way… Now!

Committed to helping you create a new life

Mind & Emotion Mastery

The Formula to Master your Mind and Emotions

Your mind and emotions are the keys to unlocking your full potential and living a happy, fulfilled, and successful life.

This online course is designed to help you learn about yourself, transform your way of thinking, gain clarity, and live from a greater level of self-awareness.

You will learn specific strategies that will enable you to overcome challenges in your personal or professional life and feel confident and inspired to achieve your goals and desires'

www.mimibland.london

Printed in Great Britain
by Amazon